CAREERS IN THE
UNITED STATES NAVY

A CAREER SHOULD BE MORE THAN JUST a job. A real career should also be a reflection of who you are and what you believe. It should be much more than putting in the hours and receiving a paycheck. People who love what they do tend to be happier and more productive than people who just go through the motions.

Since your career will be a big part of your life, it behooves you to pick something that will make your life more interesting. Something that provides you with challenges and adventures you can't get anywhere else. The fact that you are reading this report suggests that you already know this. That is why you are thinking about pursuing a career in the United States Navy.

The United States Navy's most famous recruiting slogan proclaimed, "It's not just a job, it's an adventure." The world's largest navy can take you places you couldn't possibly go on your own to do things you never knew you could do. It can send you around the world and then send you to college. It can put you on the front lines of history. Most importantly, the Navy can make you a part of something much larger than yourself.

Today's US Navy is the most powerful fleet in history. Its 280-plus ships include half of the world's aircraft carriers, two-thirds of the world's nuclear-powered submarines, and the biggest surface fleet of frigates, destroyers and cruisers on the planet. In terms of gross tonnage – the standard means to determine the size of a ship – the US Navy weighs roughly as much as the next 13 largest navies combined. Navy bases encircle the world, from the East, West and Gulf Coasts of the United States, to major installations in Italy, Japan, Spain, Singapore and Greece, plus smaller bases and piers around the world. In modern corporate-speak the Navy is what is known as a "distributed organization," because its operations are widely distributed around the world. Unlike other distributed organizations, whose operational nodes are generally in stationary buildings, the Navy's most important assets are in motion, every day. The US Navy is a unique institution that employs thousands of people with a wide variety of skills to fulfill its role as "the tip of the spear" of American military might. If you really want it, there may be a place for you.

Take careful note of the information contained in this report. You will find sections on how to get into the Navy, what kind of education you should seek out, where you are likely to be posted and how much money you can expect to earn at various stages in your career. If you like what you read here your next step should be to talk to a Navy recruiter. Walking into a recruiting station does not incur any obligation on your part, so stop by and chat with a recruiter whenever you have the chance.

YOUR ASSIGNMENT

YOU CAN GET STARTED ON YOUR NAVY career even while you are still in high school. Join an organization that teaches you something about discipline, like JROTC or Outward Bound. Study hard in class and learn everything you can about the Navy.

If your high school sponsors a Junior Reserve Officer Training Corps, join it. Even if one of the other services cosponsors it you will learn about military bearing and how to conduct yourself if you do decide to join after graduation. JROTC membership does not involve any requirement for future service. Thousands of high school students spend four years in JROTC programs and never join the military. During their high school careers, however, they spend their weekends doing things most kids don't get to do, like going on adventure trips, taking tours of military installations and learning about one of the most important institutions in history. If your school offers this opportunity, take advantage of it.

The military can be a great place for unfocused or misdirected people to get a new start in life, which may foster the impression that the military accepts recruits with below average intelligence. The truth is that the average educational level of people in the military is much higher than that of the US population in general. Additionally, only five percent of recruits in any given year can be high school dropouts, and they are required to earn General Equivalency Diplomas while they are in boot camp.

There are many ways to learn about the Navy. You can subscribe to *Navy Times,* a privately owned newspaper that covers the Navy, and join the United States Naval Institute, which publishes the magazines *Naval History* and *Proceedings.* Hollywood has been kind to the Navy, with hundreds of reasonably accurate movies to choose from that will give you a taste of what you are in for if you decide to join up.

HISTORY OF THE NAVY

NAVIES HAVE PLAYED AN IMPORTANT part in global affairs for thousands of years. The first recorded sea battle took place in 1210 BC in the waters off Cyprus, a small island in the eastern Mediterranean Sea. The sea is a precious resource. It provides food, minerals and a commercial highway to those who can control it. Nations deploy navies to protect the waters that matter most to them. In this way, navies have historically been the police of the sea, patrolling their waters the same way police departments patrol their streets. To this day, the primary function of most navies is to make sure that their citizens have safe and reliable access to the sea.

The United States Navy is actually older than the country it serves. Seeing the Revolutionary War with Britain looming on the horizon, the Continental Congress on October 13, 1775 authorized the creation of a naval force beginning with two ships. By the time the Revolutionary War began the following summer, the 13 American Colonies had established a small coastal defense force. Thoroughly outgunned by Britain's mighty Royal Navy, the world's largest for more than 300 years, the Continental Navy still managed to play an important role in the fight for American independence. John Paul Jones, the first American naval hero, made his name during the Revolutionary War by beating the British ship Serapis at the Battle of Flamborough Head. An interesting battle – look it up on the Internet.

By 1784, the Continental Navy existed only on paper. With the revolution behind them, most Americans were more concerned about building their newly independent country than they were about defending themselves from an unlikely attack. Even the Continental Army, by far the largest armed service, was reduced to a string of state militias and a tiny federal force. The European immigrants who came to America in the early years left their home countries for many reasons, one of which was the constant wars that raged across the continent. Protected by the Atlantic Ocean on one side and a vast wilderness on the other, Americans felt no need to maintain a large military. In fact, most were glad to relieve themselves of the burden.

This changed in 1794 when American shipping was routinely forced to pay bribes to the so-called Barbary Pirates, criminals who operated in the Mediterranean Sea. The new United States Congress authorized the creation of a new Navy to deal with the threat. In 1805 US Marines invaded the Libyan city of Tripoli, the pirates' home base, and forced an end to the piracy. This was the first time American forces were directly engaged in a foreign country. It also demonstrated the effectiveness of using the Navy to deploy the Marine Corps, a force specially trained to invade from the sea. To this day the Marine Corps is part of the Department of the Navy and the Navy-Marine Corps Team is deployed around the world.

The Navy played a vital role in the War of 1812. With a core of six technologically advanced frigates the US Navy routinely crippled or sank British ships in one-on-one battles. The six ships, one of which, USS Constitution, still exists and is moored in Boston Harbor today, were built using "live oak" construction. Interior structures were made from enormous single pieces of oak cut directly out of individual trees. The ships were extremely expensive but much better than anything else afloat at the time.

The American Civil War split the Navy in two, with some ships staying in the US Navy, and those based in the South being incorporated into the Confederate Navy. The Civil War saw the end of the wind-powered Navy, as both sides built steam-powered vessels clad in steel. Traditional sailing vessels continued to serve for a few decades but were replaced with steam-powered ships as they were retired.

The Navy again shrank after the Civil War. It was given a boost in the 1880s by several factors, most prominent of which was the publication of *The Influence of Seapower upon History, 1660-1783* by Alfred Thayer Mahan, a US Navy captain. In simple terms, Mahan argued that nations that can control important sea-lanes can control the commerce that uses them. In this way, they can strengthen allies, weaken enemies, and enjoy prosperity for themselves. Mahan's theories are still required reading for US naval officers today.

The boost came along at an opportune time. When the warship USS Maine mysteriously exploded off Havana, Cuba, the United States in 1898 found itself at war with Spain, which then owned Cuba. The US Navy won the Spanish-American War, decisively beating a European fleet and encouraging newly elected President Theodore Roosevelt to push for a new fleet, known as the Great White Fleet, upon taking office in 1901. Although impressive ships individually, the fleet of 16 battleships was no match for the major navies of the day, especially those of Britain and Germany. Starting in 1917, the Navy played a small role in World War I, mostly escorting troop ships across the Atlantic and patrolling the American coast. As soon as the war was over the Navy shrank again.

The modern Navy was born in the period between the wars. Limited by treaties signed after World War I, major powers, including the United States, could not build very many battleships. The enormous, expensive ships bristling with guns were the most fearsome expressions of a nation's ability to project power around the world, so each of the world's major navies was only allowed a limited number. Although intended to reduce the possibility of another world war, the treaties mostly encouraged navies to look into alternatives like aircraft carriers.

The Navy launched its first aircraft carrier in 1922. The USS Langley was a coal-carrying ship reconfigured to carry aircraft by adding a huge wooden flight deck on top. A ridiculous sight to modern eyes, the Langley's flight deck looked more like a roof than a runway but it worked well enough to prove that planes could be flown safely and effectively from ships. The combination allowed the Navy to take aircraft anywhere in the world, without the need to build runways or ask foreign governments for permission to base aircraft.

The Navy also invested heavily in its submarine force during this period. Used most effectively by the Germans in World War I, submarines had been embraced reluctantly by the US Navy in part because they were a new and mysterious technology, and in part because many tradition bound naval leaders saw them as furtive and therefore fundamentally dishonorable. German success at disrupting transatlantic commerce forced a change of heart.

The Japanese attack on the US Navy base at Pearl Harbor, Hawaii on December 7, 1941 propelled the Navy from a medium-sized force to the largest navy in the world by 1945. The Navy played a huge role in World War II, fighting simultaneously in both the Atlantic and Pacific theaters of war. The Navy built and launched thousands of vessels during the war, many of which were sunk or damaged beyond repair. By war's end, however, the US Navy had eclipsed the Royal Navy as the world's largest and most powerful. It has remained so ever since.

The US Navy again shrank after World War II, but not by as much as it had after earlier wars. Tension with the Soviet Union led to a state of affairs known as the Cold War, in which both sides maintained large militaries on high alert. The Navy was critical during this period. In its role as the police officer of the world's sea-lanes, the Navy made sure that commerce kept flowing to friendly countries around the world. Then as now, more than 90 percent of the world's trade moves by sea. The Navy developed nuclear-powered vessels during the Cold War in part because with no need to refuel they could stay at sea longer. This deterrent presence continues to be important today.

Conflicts in Korea and Vietnam were key moments during the Cold War, and the Navy stepped up for both. In each case the Navy brought to bear classic naval capabilities. First, the Navy used its superior firepower to block commerce into and out of Korea and Vietnam. In addition, the Navy's aircraft carriers provided valuable air cover to ground troops without the need to base aircraft on land.

Computers have revolutionized the Navy just as they have revolutionized most other enterprises. Today's naval vessels are floating computer labs, bristling with electronics that keep an eye on the entire world. Missiles are targeted using satellites linked to layers of radar. Satellites also keep ships in touch with the rest of the fleet and even allow for more mundane functions like sending personal emails to friends and family back home. Ironically, all US military satellites are owned and operated by the US Air Force but the Navy uses about 85 percent of their bandwidth because of its global distribution.

Today's Navy patrols the world's hot spots all day, every day. The Navy is known as the "tip of the spear" because it has been the first to respond to conflicts all over the world. Its unique ability to take up station near conflict zones without the need to ask another government for permission makes it critical in times of crisis. Naval vessels have been stationed in the Persian Gulf since the 1991 conflict in Iraq, providing security for oil tankers and other cargo vessels conducting their business there. The Navy-Marine Corps Team fired the opening salvos against Afghanistan after the terrorist attacks on the United States on September 11, 2001. Quite an impressive feat when you consider that Afghanistan is more than 300 miles from the sea.

Today's fleet of about 281 ships is the smallest since before World War II. Individual ships are so much more powerful than their predecessors, however, that the smaller Navy still packs more punch than ever before. Plans call for building the fleet up to 332 ships by 2025, an achievable but optimistic goal. Sophisticated naval vessels are extremely expensive and take many years to build, making them easy targets for politicians looking to shave money from the military budget. If the Army asks for $2 billion to buy tanks but only gets $1 billion, it will just buy half as many new tanks. But if the Navy asks for $2 billion – the cost of a single destroyer – and gets only $1 billion, it cannot build this ship. This is a problem that plagues all navies. The Navy, Marine Corps and Coast Guard have pledged to work more closely when acquiring systems and planning strategies, in part to make their budgets go farther. A Cooperative Strategy for 21st Century Seapower, commonly known as the Maritime Strategy, plots a bright future for the nation's sea services Read all about it here: www.navy.mil/maritime/MaritimeStrategy.pdf

WHERE YOU WILL WORK

THE NAVY CAN SEND YOU JUST ABOUT anywhere in the world. For thousands of years navies have been plying the seas, taking their sailors to the farthest reaches of the world. Navies have to establish bases, or at least friendly relationships, everywhere they go, in order to be able to pull into port, take on fuel and food, and conduct repairs. If you join the Navy you will, indeed, see the world.

The military likes acronyms. The two that apply to this section are "CONUS" and "OCONUS." CONUS stands for "continental United States." OCONUS stands for "outside the continental United States," which includes Alaska and Hawaii. The Navy's biggest installations in-CONUS, known as fleet concentration areas, are in Norfolk, Virginia and San Diego, California. If you serve long enough you will undoubtedly be stationed in at least one of these locations. Other major installations can be found in Mayport and Pensacola, Florida; Newport, Rhode Island; Bremerton, Washington; New London, Connecticut; Great Lakes, Illinois; Gulfport, Mississippi; Corpus Christi, Texas, and all around the Washington, DC metropolitan area. Naval personnel are also assigned to hundreds of smaller facilities and recruiting stations across the country.

Pearl Harbor, Hawaii is easily the best-known OCONUS Navy installation. The Navy has no bases in Alaska but routinely assigns personnel to small military facilities there and to bases operated by the other services. On foreign soil, the Navy has major installations in Bahrain, Diego Garcia, Djibouti, Greece, Italy, Japan, Singapore and Spain. Naval vessels pull into ports all over the world for days or weeks at a time to take on fuel and food and give the ship's company some time off.

Naval personnel are also assigned to joint and combined operations. Joint, in military terms, refers to operations undertaken by multiple services. Combined operations are those undertaken by militaries from multiple countries. That is why there are naval personnel serving on US Army bases in Stuttgart,

Germany, which is a long way from salt water, and aboard British Royal Navy vessels, among other examples.

Naval personnel are typically reassigned every three years or so. That doesn't include deployments, which generally last six to eight months, involuntary orders to war zones, or short-term orders for assignments like a year of graduate school. If you stay in the service long enough you could live in a dozen different places and visit hundreds more.

YOUR WORK DUTIES IN THE NAVY

THE NAVY OFFERS HUNDREDS OF career paths, all of which can easily provide enough challenges to keep anybody interested for 20 years or more. Most skill sets have enlisted and officer career paths, and some also have an intermediate path for warrant officers. Duties and responsibilities are very different for enlisted sailors, warrant officers and commissioned officers.

Enlisted personnel do the work of the Navy.

Enlisted personnel have the skills to keep the ships moving, process the interminable paperwork, guard the bases, fix the computers and attend to all the other tasks necessary to keep such a large organization up and running. Specific career paths are known as ratings. About 85 percent of Navy personnel are enlisted.

Senior enlisted personnel, known as chief petty officers, are the Navy's middle managers. They wear the same uniforms commissioned officers do, but with different insignia denoting their rank. Chiefs are critical to the functioning of the Navy. They are the intermediaries between the officers who make the big decisions and the junior enlisted personnel who carry them out. It has been said that "chiefs run the Navy." While not technically true, it is undeniably true that very little happens without chiefs moving things along.

Warrant officers are an intermediate grade between enlisted sailors and commissioned officers.

The Navy has relatively few warrant officers; only about 3,500 out of a total active-duty strength of 332,000. That is a bit more than one percent of naval personnel. Warrant officers are senior technical specialists within their fields. Most ratings and designators, it should be noted, do not have warrant officer career paths. Those that do, like intelligence, cryptology and information technology, tend to be technical specialties. Nobody becomes a warrant officer right away. To apply for a commission as a warrant officer an applicant must already be a chief petty officer or senior chief petty officer and have at least 12 years of service. Warrant officers may be outranked by commissioned officers but smart commissioned officers generally give them the latitude to do what they do best.

Commissioned officers, usually known simply as officers, are the Navy's leaders.

A commission is a standing order from the commander in chief – the President of the United States – to act on his behalf. Commissions are rooted in thousands of years of tradition and can trace their origins to the days when kings tapped exceptional soldiers on the shoulders with their sword and dubbed them knights, empowered to enlist soldiers to fight wars on the king's behalf. Officers are legally responsible for all of the actions taken by the Navy, starting with the people directly beneath them in their individual chain of command. It is routine for 22-year-old ensigns, newly graduated from Officer Candidate School (OCS) to immediately be put in charge of dozens of junior enlisted personnel. This is a heavy responsibility, and one you should bear in mind if you want to apply for a commission.

Job classifications for enlisted personnel are known as ratings.

Ratings are combined with rates to create formal titles. For example, an intelligence specialist is known by the rating abbreviation "IS." An intelligence specialist with the rate – synonymous with "rank" – of petty officer, second class would be known as IS2, as in "IS2 Smith." When promoted to petty officer, first class IS2 Smith would be known as IS1 Smith. Chief petty officers, senior chiefs and master chiefs have similar abbreviations added to their titles: ISC, ISCS and ISCM, for example.

Enlisted personnel can also earn Navy Enlisted Classifications, or NECs, that correspond to their ratings. Intelligence specialists can earn NECs in Naval Imagery Interpretation, Naval Special Warfare, Strike Planning and Operational Intelligence. Intelligence specialists can earn just one NEC, all four or something in-between. This basic arrangement applies to all ratings.

Enlisted personnel can also earn warfare qualifications to enhance their ability to operate within larger communities. The most common warfare qualifications are for surface, subsurface and air warfare. Sailors who earn warfare qualifications are entitled to wear special pins on their uniforms. Sailors who have served aboard aircraft carriers are often able to earn both surface and air warfare qualifications in the same tour and are said to be "double-pinned." The system of NECs and warfare qualifications allows sailors to tailor their individual careers with great precision.

The number of individual ratings varies over time. Ratings are routinely created, eliminated and merged in order to meet emerging needs. Some ratings temporarily close to new accessions because they have too many people, while others actively recruit because they have too few. Only a recruiter can give you up-to-the-minute information about billets available in specific ratings.

The full list of ratings currently available to aspiring sailors is available at www.navy.mil. Here is a small sampling:

Air-traffic controller

Aerographer's mate

Aviation warfare systems operator

Builder

Construction mechanic

Engineering aide

Steelworker

Boatswain's mate

Culinary specialist

Fire control man

Hospital corpsman

Master-at-arms

Mass communications specialist

Musician

Postal clerk

Religious programs specialist

Special warfare operator

Damage control man

Hull maintenance technician

Navy diver

You should study the full list. You may be amazed at the career specialties you can pursue in the Navy. Culinary specialists, for example, get part of their training at the Culinary Institute of America in New York, one of the country's finest culinary schools.

Job classifications for officers are known as designators.

All officers are first assigned to one of four communities: restricted or unrestricted line officer, staff officer, limited duty officer or warrant officer. In simplest terms, unrestricted line officers are those officers in front-line Navy fields like surface warfare, subsurface warfare and aviation. Restricted line officers include those assigned to functional areas like intelligence, oceanography and public affairs. The primary difference between restricted and unrestricted line officers is that only unrestricted line officers are eligible to command ships and aircraft squadrons. Staff officers are essentially everybody else, from physicians and nurses to chaplains and lawyers. Limited duty officers, or LDOs, are former enlisted personnel commissioned based on their performance whether or not they have earned a bachelor's degree. They are limited only in that they generally cannot transfer into other specialties. Warrant officers are technical specialists who had to achieve the rank of chief petty officer in order to apply for a warrant commission. All officers of equal rank have the same authority and privileges no matter what their commissioning source.

The main designator communities are aerospace maintenance, naval aviator (pilot), naval aviator (naval flight officer), chaplain, civil engineer, cryptology, engineering, intelligence, judge advocate general (lawyer), medical services corps, naval reactors engineer, nuclear officer (surface), nuclear officer (submarines), nurse corps, oceanography, public affairs, special warfare, special operations, supply, and surface warfare. Like ratings, designators are always in flux. Check with an officer recruiter for the latest news.

TRUE TALES FROM THE NAVY

I Am a Surface Warfare Officer

"My designator says I'm a surface warfare officer, or SWO. I'm also known as a ship-driver. I like to think of myself and my fellow SWOs as being the backbone of the Navy. The chiefs will disagree because they think they're the backbone of the Navy. That's their prerogative but as I see it the Navy is about ships and SWOs command them.

I am a true Navy professional. I set my sights on going to the Naval Academy when I was a freshman in high school. You have to begin preparing early. The entrance requirements are very high; in the same league with a top-tier private university. Only about 10 percent of applicants are selected. The academy looks at your grades, civic involvement, physical fitness and other attributes like foreign-language skills. It also looks at your moral character. That's hard to define with any precision, but if you've ever been arrested you can forget about getting into the academy. You have to have the whole package in order to make the cut.

The academy was hard. Unlike most civilian schools, academy cadets are usually confined to the campus and have to abide by very strict rules of conduct. Every day starts with a workout. There is no such thing as summer vacation. There aren't any part-time jobs, either. Cadets are paid a small stipend so they can concentrate on their studies. Cadets who don't make the grade are ruthlessly cut. About a third don't make it to graduation. It's a very prestigious education, to be sure, but it's not for everybody. In fact, it's not for most people.

Cadets are commissioned immediately upon graduation. After a week or so of leave, it's off to specialized training and your first assignment in the fleet. It's a bit intimidating, but it's also very exciting. I breezed through SWO school and was assigned to a frigate, a small surface ship, as a division officer, or DIVO. As a DIVO I was in charge of a clutch of junior enlisted personnel. I did their paperwork, kept them out of trouble and provided something like leadership. It was a tough job.

It was an important job, too. Being a junior officer is a little intimidating. I was 22 years old. Many of the junior enlisted personnel who reported to me were much older than I was and knew more about the Navy and the ship, but they still had to take orders from me. I was lucky to have a good chief petty officer on my team. Chiefs, the most-senior enlisted personnel, are the Navy's middle managers. Even though I outranked my chief, my chief guided me and provided me with extremely valuable mentorship. The nature of this relationship is very difficult to explain to people who haven't experienced it. If you think the military is all about people of high rank blindly ordering around people of low rank, you are dead wrong. To this day, I consult my chiefs before I make any major decisions.

My career has followed the normal SWO path. I have been assigned to positions of increasing responsibility in several sea and shore billets. I am currently an O-5, or commander, and the executive officer, or XO, of a guided-missile cruiser, the largest type of surface combatant. Along the way I have earned two master's degrees: one at a civilian university and one at the

Naval War College. The Navy paid for both of them. They'll come in handy when I retire and look for a job in the civilian world.

This is a feast-or-famine kind of job. The Navy calls it the 'sea-shore rotation.' That means that I spend a two-to-three-year hitch on shore duty followed by a two-to-three-year hitch on sea duty, during which I might be underway half of the time. Shore duty comes with regular hours but little adventure. Sea duty comes with irregular hours and lots of adventure. I have a spouse and children, so I like to spend time at home. But I'm a SWO at heart, and nothing can replace the thrill of life at sea. This, more than anything, is the sacrifice the Navy demands of all its members."

I Am an Interpreter

"My rating is 'cryptologic technician, interpretive.' In the civilian world I would be known as a translator. Just for the record, I love my job. I get to travel around the world using my skills to help Navy personnel communicate. It's an important contribution.

I enlisted in the Navy right out of high school. I had no idea what I wanted to do but I did pretty well on the ASVAB so I had my pick of essentially every rating in the Navy. The ASVAB (Armed Services Vocational Aptitude Battery) is an aptitude test developed and maintained by the Department of Defense. Just for kicks, I also took the Defense Language Institute's aptitude test, and I did well on that, too. So the Navy sent me to Monterey, California for six months to learn French. If six months doesn't seem like a long time to learn a language I can't argue with you. DLI makes it work, however, by immersing students in the target language for all six months. Classes are eight hours per day, five days per week. It is against the rules to speak English in the hallways or the cafeteria between classes. Students are required to speak their target language even in their own dorm rooms. It works. Your brain sometimes feels like it's going to explode, but it works.

As a CTI, I am responsible for translating documents and providing simultaneous interpretation. That means I get to tag along with senior officers and translate for them when they do business with French speakers. I also stand at the front of the room and translate for instructors when they deliver training in French-speaking countries.

There are many French-speaking countries. French is widely spoken in Africa and the Middle East. In recent years I have spent more time working in Africa than in Europe, which is really cool. I spent three months in Congo, for example, and have been on short trips to Gabon, Cameroon, Morocco, Mauritania and Algeria. I would never go to any of these places on my own. As a CTI, the Navy pays me to go there – one of many reasons to love this job.

I am currently a first-class petty officer, or CTI1. I'm working on a bachelor's degree in French and hope to become a chief someday. I really like what I do and want to manage the activities of CTIs in support of larger projects. I have some ideas that I think would really improve the process. As a chief, I'll be able to carry them out.

I'd recommend this job to anybody with a flair for languages. The Navy needs translators and interpreters for many different languages, and is willing to pay a bonus for some. I get an extra $300 per month to maintain my fluency in French. Speakers of non-Western languages like Chinese, Farsi and Arabic get a $600 monthly bonus. If you have the skills you won't be bored. In fact, that's what I like best about this job. The Navy values my skills and rewards me with challenging work."

I Am a Public Affairs Officer

"I have one of the most important jobs in the Navy. As a public affairs officer, or PAO, I am responsible for making sure that the Navy gets its messages across to all of its audiences. Naval personnel, their families and civilian employees are our internal audiences. Our external audiences include the media, foreign militaries and citizens, and, most important of all, American citizens. We work for American citizens, and they deserve to know what we're up to.

I stumbled into public affairs almost by accident. I enlisted in the Navy after earning a bachelor's degree in mathematics. I could have applied for a commission but wasn't interested. I knew I needed some serious experience before I took on the added responsibility of leading people. So I became an engineering aide, or EA, and worked for a succession of Construction Battalion units. These are the famous Seabee units, the battlefield engineers whose motto is 'We build, we fight.' I was a 'Bee for 10 years and loved every minute.

For the last couple of years in the 'Bees, however, I was assigned to recruiting duty. I loved that, too. I dealt with media representatives, planned events, and gave presentations. So I put in a package to become an officer. I was selected, went to Officer Candidate School and was assigned to the News Desk at the

Pentagon. First, OCS was much harder than boot camp. Second, being one of dozens of junior officers on the News Desk was the most frantic job I have ever had.

The PAO community is the smallest officer community in the Navy. There are only about 200 of us. We are assigned to bases all over the world and to aircraft carriers at sea. PAOs assigned to shore billets routinely deploy on ships for a few weeks or months to support important operations. We have a very diverse portfolio of responsibilities, which is one of the things I like best about the job.

For example, I run all the base newspapers and websites. They are the primary means of communicating with our internal audiences. This is challenging because rumors spread very rapidly in small, closed communities like ours. Issues like base maintenance, ship's schedules and changes in allowances have to be communicated effectively in order to prevent confusion. I am also the primary public affairs advisor for my boss, an admiral in charge of several bases. I give the boss tips on talking to the media and to our own people. My expertise helps keep everybody accurate and on message.

My other main responsibility is communicating with our external audiences. In practical terms this means developing good working relationships with the media. This can also be very challenging. The military is a complex institution. Reporters really need to take the time to understand an issue before they put it in the newspaper or on television. I help them to understand the Navy and how we do our business so they can present an accurate picture to the American people. The bottom line is that everything we do, we do with the people's money and, most importantly, their moral authority.

I am currently a lieutenant commander. I'm up for promotion to commander in a few months and I'm pretty sure I'll make it. My community is so small that I know all of my competitors. An honest assessment says that my odds are excellent. I'm already past 20 years in the service, but I think I'll stay in for a few more. Through the Navy I've lived in Washington, San Diego, Boston, San Francisco, New York, Chicago, and Naples, Italy, and served aboard ships in the Persian Gulf and the South Pacific. The Navy paid for me to get a master's degree from a civilian university and gave me a year off to complete it. I like the nature of the work and I take real pride in being in the Navy. Even though I could retire today, I can't think of a reason why I should."

I Am an Intelligence Specialist

"I got into this career in a very roundabout way. I didn't join right out of high school or college. In fact, I didn't join until I was in my early 30s. I am a journalist by trade. I used to cover the Navy. One day it occurred to me that the people I was covering always seemed to be having a better time in their work than I was. So I joined the Navy Reserve as an intelligence specialist.

Intelligence has more in common with journalism than you probably think. Both jobs are about analyzing a large amount of information and turning it all into a small amount of intelligence. That's what intelligence is: information subject to analysis. As a journalist, I regularly spent all day doing interviews and background research just so I could write a story that could be read in five minutes. As an intelligence specialist, I do essentially the same thing but the resulting "story" is usually classified. Intelligence has turned out to be a fascinating way to use skills I already had.

Not that I had all of the skills I would need. Journalism gave me good analytical skills but didn't give me the specifics of the intelligence business. I had to attend Operations Specialist "A" School in Dam Neck, Virginia to learn those. I can't reveal very much about what I learned there, but it was definitely time well spent.

In order to qualify for this rating I had to get a high score on the ASVAB and qualify for a security clearance. It took about a month just to do the paperwork for the clearance. Then it took a year for the background investigation. Investigators got in touch with my family, friends and former employers, took a look at my financial history and checked out a list of every time I had traveled outside the United States. It was very thorough. If you have credit problems or a police record you can forget about getting a security clearance. You must be a US citizen in order to be granted a security clearance. There are many other restrictions.

A security clearance is a great privilege, and not something to be taken lightly. I know things that could be valuable to those who would like to do us harm. I can only talk about them with people who have the right clearance, only in appropriately secure spaces and only if they need to know. The process is admittedly cumbersome, but it works.

As a Reservist I have been able to explore several different areas. I spent a few months in Hawaii doing imagery analysis, which was fascinating. I also served as petty officer of the watch on an operational intelligence watch floor in Naples, Italy for a while. My regular Reserve unit does background research to prepare trip reports for officers preparing to travel. They need detailed information about their destinations and we provide it to them. I have worked my way up to IS1 in this unit and will be eligible to take the chief's exam later this year. I'm already studying.

I like being a Reservist because I get the best of both worlds. I still work as a journalist in civilian life but I get to delve into the world of military intelligence at least one weekend per month and two weeks per year. Over the years I have volunteered for longer stints on active duty and have been involuntarily recalled once. Those six months I spent in a combat zone weren't exactly fun but they were very illuminating. If called upon I would do it again. It's what I signed up for."

PERSONAL QUALIFICATIONS

NOT JUST ANYBODY WILL BE HAPPY IN THE NAVY. IT CAN BE A VERY CHALLENGING WAY OF life. If you are serious about pursuing a career in the service you should be, first and foremost, dedicated to the idea of service. After that, it definitely helps to be patient and to be able to always keep the big picture in mind.

"Service" is not just a handy word to describe employment in the armed forces. Military personnel, known as service members, really do serve the people and interests of the United States of America. If you think joining the military is all about your personal glory, please look into another line of work. The military will demand great accomplishments from you; accomplishments in which you will be able to take justifiable pride. But service is never about you. Service is about defending the American way of life. Some service members complain about the fact that so few Americans ever serve in the military, and that most Americans take our military might for granted. One percent of Americans make it possible for the other 99 percent to live in peace and freedom.

Much of the military is about "hurry up and wait." All of the services are very structured, hierarchical and bureaucratic. If you are assigned to deploy on a ship, for example, you will be required to prepare – get a physical, process sea-duty orders, update your will and attend to a thousand other little things, all of which will have to be finished before the petty officer on the pier will let you board the ship. Everybody else assigned to the same deployment will have to go through the same checklist. A battle group consists of almost 10,000 people. When 10,000 people have to go to the same workplace to do the same things, all pretty much at the same time, the lines can get long. As confused and inefficient as it seems, the military system usually works – eventually.

Successful service members have to keep the big picture in mind at all times. The bureaucracy may drive you crazy, the hours can be relentless, the work can be dangerous and the person in charge – the commander-in-chief – may not always be the person you voted for. None of that will matter. Militaries have been one of the driving forces of statecraft for as long as there have been states. They have always been bureaucratic and unable to make big changes as quickly as anybody would like, and they often report to civilian leaders who are not very popular with the rank and file. Read up on military history. It helps to put your own misgivings into perspective when you realize that John Paul Jones and Bull Halsey had to grapple with the same issues.

ATTRACTIVE FEATURES

THE NAVY HAS BEEN ONE OF THE largest employers in the United States for more than 200 years for a reason. It is a great place to work. You will have adventures you could not possibly have anywhere else. You will also be on the receiving end of one of the world's greatest benefits packages. Ultimately, you will gain great satisfaction from serving your country.

In the 21st century most Americans rarely venture very far from their desks. Adventure is limited to weekends and vacations, and usually is expensive. The Navy will send you on adventures you can hardly imagine, for months or years at a time, and pay your way. Your civilian friends can marvel at the precision of aircraft-carrier operations when they see them on television, but you can serve aboard a carrier and be a part of making it happen. Want to go to Europe, Central Asia, South America, Japan and Africa? Depending upon your exact job, you may go to all of these places and more in a matter of months. As the nature of warfare and the threats to the United States evolve, the US military is making new connections in faraway corners of the world. More military personnel are currently deployed to Africa than at any time in history.

In exchange for volunteering, the Navy will provide very generous benefits, including 30 days of paid leave per year, time during the day to go to the gym and stay in shape and access to commissaries and exchanges, subsidized grocery and department stores reserved exclusively for service members and their families. The healthcare plan is excellent. The Navy owns the hospitals and clinics, and employs the doctors required to take care of you. The GI Bill can pay for a college degree and the VA Loan will help you to buy a home. Service members can also collect a lifetime pension equal to half of their basic pay after serving for only 20 years. That means you could retire at 38! The compensation package is hard to beat.

Military service is very satisfying. Lack of participation is one of the common laments of modern life. Employees have to be told to organize the annual company picnic because no one volunteers. Charities have no problem raising money but cannot find people willing to give a few hours a week of their time. Voter turnout barely tops 70 percent for national elections and might not break 20 percent for local elections. These are all signs of the malaise that affects industrialized societies in which most people are generations removed from real hardship.

Modern life is so easy and routine, it can become boring and meaningless. In the Navy, you won't have this problem. You will be one of the people who stepped up to the challenge posed by a dangerous and uncertain world. You will take great pride in that.

UNATTRACTIVE FEATURES

IT SHOULD COME AS NO SURPRISE THAT Navy life can be difficult. While you are on active duty you "belong" to the Navy in a way that no other employer could get away with.

Adventures are great in the short term. Who wouldn't want to take advantage of an opportunity to become familiar with the inner workings of an aircraft carrier or go on a port call in Africa? But anything can become a grind if you do it long enough. Deploying on an aircraft carrier for six to eight months, for example, is positively thrilling at first. Then the day comes when you realize that you are thousands of miles from home, bouncing around on the waves in a big tin can. This is especially difficult for naval personnel with spouses and children. Adventures can also be dangerous. Ultimately, your job is about being prepared to kill enemies and destroy property in order to protect the people and lives that are important to you. An adventure in Africa, or any other exotic place, can go horribly wrong if you're bitten by the wrong mosquito, drink some of the local water, or get a bomb hurled into your quarters. Adventures are inherently risky. You may be seriously injured, or killed in action.

All service members are subject to a law known as the Uniform Code of Military Justice, or UCMJ. Under the UCMJ all active duty service members essentially belong to their service. You may usually work during the day but evenings and weekends do not belong to you. They are "liberty" and can be turned into working hours at any time. There is no overtime. The Navy will control your private life to a degree no other employer ever could, restricting your political activities and associations, requiring you to file an itinerary whenever you travel, and subjecting you to random drug testing. You can be ordered to get on a ship and sail away from your loved ones on a few hours' notice. The military is not like civilian society. Service members are considered national assets and must be ready to be used at all times.

You get to know your fellow service members in a way that other employees never do. You will live with them, go to sea with them, have adventures with them, share secrets with them, go to weddings and funerals with them. The Navy is a very cohesive society. Many people find this to be comforting. There is no doubt, however, that you will have to give up some of your freedoms and privacy in order to become a part of it.

Nobody likes the military bureaucracy. Everything it does takes too long and costs too much. It's so big it probably can't be fixed. The Pentagon, the military's headquarters in Washington, is literally the biggest building in the world. People on one side of the building don't even know how to find people on the other side, let alone what they've been up to lately. Rules cannot be broken, but they're so complex that everybody bends them, and in differing and inconsistent ways that make it very difficult to get anything done outside your immediate department. If you are a conscientious person, this will bother you. You will have to learn to tolerate and function within the system.

EDUCATION AND TRAINING

YOU WILL RECEIVE EXTENSIVE TRAINING after you get into the Navy. In fact, the only time you won't be in some kind of training is when you are deployed to a combat zone. If you want to succeed in the Navy, you will have to like formal learning on an ongoing basis. Additionally, how much education and training you have when you join will have a significant effect on your career advancement.

If your highest academic credential is a high school diploma you can join the Navy as an enlisted service member. If you have at least a bachelor's degree you can apply for a commission to become an officer. Although officers outrank enlisted personnel, do not leap to the conclusion that being an officer is inherently better than being enlisted. Enlisted personnel are the Navy's technical specialists – they do things. Officers are leaders – they establish policy and direct the efforts of enlisted sailors. Officers get paid more but they also assume a much greater degree of legal and moral responsibility for their actions and the actions of those under their command. If you want to become an expert hull technician, photographer or special forces commando, among many other choices, you should enlist. If you want to spend your time managing projects and directing the activities of others, set your sights on becoming an officer.

If you enlist straight out of high school you will have opportunities to apply for a commission later in your career. The maximum age to earn a commission for most designators is 35, although some allow commissions up to age 42. If you want to be an officer but cannot manage a bachelor's degree before joining, the Navy will help you earn one. You can apply for a commission then.

If you intend to enlist, you should work to your maximum potential in high school. Math and English are especially important. History will help too, as will working knowledge of a foreign language. Take gym class as seriously as the rest of your classes. You should be able to comfortably run three miles and do at least 50 push-ups and sit-ups in order to be prepared for boot camp. Boot camp isn't easy, but don't let an irrational fear of it put you off joining the Navy. You will not be harangued by crazed drill instructors or pushed to within an inch of your life. Modern boot camp is a largely intellectual exercise conducted mostly in classrooms. The physical requirements are challenging, but they aren't superhuman. After the first couple of weeks, in fact, most new recruits start to enjoy the structure and challenge.

If you intend to apply for a commission you will need to earn a bachelor's degree before joining. Almost any major will help you to earn a commission, although some are in greater demand than others. Sciences, engineering and foreign studies are in the greatest demand. Good grades will also be essential. Earning a bachelor's degree is no guarantee of being awarded a commission. It just means that the officer recruiter will let you do the paperwork to apply for one. The Navy does almost no officer recruiting because it doesn't have to. Only a small number of applicants are chosen.

If you choose to attend a civilian university you can either enroll in the Navy Reserve Officer Training Corps, or ROTC, or apply for a commission after you graduate. ROTC is a minor in military science and comes with a series of classes to get students up to speed on becoming naval officers. Most ROTC students receive some kind of scholarship. In return for the financial assistance, ROTC students spend several weekends during the school year and several weeks each summer undergoing Navy training. ROTC students are obligated to serve for at least five years after graduation. Go to this website for a list of colleges with Naval ROTC programs:

https://www.nrotc.navy.mil/colleges.cfm

If your school does not have an ROTC program, you can apply for a commission after you graduate. If you are accepted, you will go through 13 weeks of training at Officer Candidate School, the officer equivalent of boot camp.

The other option for earning a commission is to attend the US Naval Academy in Annapolis, Maryland. The academy accepts only about 1,200 students per year. Entrance requirements are very tough. Your grades, classes, extracurricular activities and volunteer work will all be taken into account when you apply to the academy. You will also need a nomination from your US senator or representative. If you are accepted, you will receive an excellent undergraduate education and be obligated to serve for at least five years after you graduate. Additional details can be found on the academy's website.

After you have completed ROTC, boot camp or Officer Candidate School, you will go on to a specialized school to learn a specific skill. Enlisted sailors go to "A" School, while officers go to whatever training program is specified by their designator. After all initial accession training has been completed both officers and enlisted personnel report to their first duty stations. No matter which path you take you will be in and out of training for the rest of your career.

The Navy provides many opportunities to get additional academic training after you join. The GI Bill will pay for some or all of your academic or vocational expenses if you agree to serve for at least five years. The Tuition Assistance program will help you to earn an advanced degree while you serve. The Navy runs its own graduate school, the Naval War College, and also sends officers to joint-service schools like the National War College and the Joint Military Intelligence College. A master's degree is required in order to be promoted to the rank of commander, or O-5, in many designators. If that is the case, the Navy will help you to meet the requirement.

EARNINGS

NOBODY GETS RICH IN THE MILITARY but everybody does well financially. The compensation package is designed to make service members secure in the knowledge that they, their families and their futures are being looked after. It is very difficult to get people to go to sea for months at a time if they feel their loved ones won't be able to get by. You will never go hungry in the Navy.

In all military services your paycheck is actually composed of several different kinds of compensation. The largest portion of your paycheck is base pay. This is your salary without benefits. In 2009 a new enlisted sailor in the rank of E-1 (seaman recruit) was paid $1,400 per month, or $16,800 per year. This may not sound like much, but keep in mind that most E-1s are in boot camp where they also get a roof over their head, three meals a day and almost no opportunities to spend money. Base pay jumps to $1,569 per month for E-2s (seaman apprentices) and $1,588 for E-3s (seamen).

As a young sailor or officer most of your needs will be

provided for, so your smallish paycheck will actually go pretty far. When you rise to E-4 (petty officer third class) your base pay will jump to $1,921 per month if you have been in for at least two years or $2,025 per month if you have been in for three years. An E-5 (petty officer second class) with four years in earned $2,335 per month in 2009. After that, your base pay will increase every two years all the way up to $5,928 per month for an E-9 (master chief petty officer) with at least 26 years of service. The highest-ranking enlisted personnel can, under certain circumstances, serve for as long as 40 years. With a combination of hard work and luck you could top out the pay scale at $6,830 per month in base pay for an E-9 with at least 38 years of service.

At some point you will probably move off base and into a house or apartment of your own. When you do, you will be entitled to a housing allowance. Housing allowances are calculated based on your rank, location and whether or not you have dependents. The housing allowance for a single E-5 sailor living in a rural area, for example, would be substantially less than the housing allowance for a married E-5 sailor living in an expensive metropolitan area. Housing allowances for some areas, especially overseas, can be as large as or larger than your base pay. Typically, however, a housing allowance will account for about one third of your paycheck.

After your base pay and housing allowance, there are additional entitlements. The cost of living allowance, or COLA, adds to your base pay when you are assigned to an area with a high cost of living. Hazardous duty pay adds to your paycheck when you deploy to a dangerous area. Your paycheck will be tax-free if you are deployed to an area declared a combat zone by Congress. That is a substantial raise. Sailors who can prove their fluency in certain languages may be eligible for a monthly bonus. The list of ways to boost your paycheck is quite lengthy. Nothing comes for free, however. All of these bonuses will require hard work on your part. Want to earn an extra $600 per month? Learn Arabic.

Pay scales for officers are higher than those for enlisted sailors. Most other allowances are higher too. Base pay for a new O-1 (ensign) starts at $2,655 per month and rises to $3,483 for an O-2 (lieutenant junior grade) with two years of service. From there the pay scale rises to a maximum of $18,061 per month for an O-10 (admiral) with at least 38 years of service. Keep in mind, however, that there are only eight four-star admirals serving at any given time.

The Navy also fields about 3,500 warrant officers, an intermediate rank that can only be attained after you have achieved the rank of chief petty officer. Nobody becomes a warrant officer by being promoted. If you want to be a warrant officer you will have to apply for a position. Warrant officers are technical experts who devote their careers to a specific technical specialty. In the chain of command they are above all enlisted personnel but below all commissioned officers. Not all designators have warrant career paths. Base pay starts at $4,037 per month for a CW-2 (the CW-1 rank is no longer in use) with at least 12 years of service, the minimum time in service needed to apply for the warrant program. A CW-3, or chief warrant officer three, the next step in the chain, earns $4,677 per month with at least 14 years of service. A CW-5 with at least 38 years of service tops out the pay scale at $8,514 per month.

OPPORTUNITIES

YOU CAN GET A RUNNING START ON your Navy career before you actually join. In addition to enrolling in JROTC or ROTC programs to learn about the military way of life you can seek out other leadership opportunities or join the US Navy Reserve.

There are more than 122,000 citizen sailors serving in the US Navy Reserve, a critical organization that provides extra personnel to the active duty Navy in time of need. Many Reservists join when they leave active duty. Sometimes they do it because they want to maintain the option to return to active duty easily, sometimes they do it to get to 20 years of service and qualify for the retirement package, and sometimes they just can't bear to completely separate themselves from the Navy.

Many other Reservists, however, join the Reserve straight from civilian life. The Navy Reserve prides itself on attracting older recruits who already have valuable skills needed by the Navy. The average age of active-duty boot camp recruits is 19, but the average age of their Reserve counterparts is 30. If you are unsure about an active-duty career you can answer many questions by doing a hitch in the Reserve. The minimum requirement is to serve one weekend per month and two weeks per year. Keep in mind, however, that you can be called to active duty at any time. You can also volunteer for temporary active duty. Special projects often come up that need to be filled for as little as 30 days. Many Reservists carve out exciting second careers in the Navy by volunteering for active duty whenever they can make the time.

Britain's World War II Prime Minister Winston Churchill, who previously served as that country's first sea lord, the equivalent of the American chief of naval operations, called military reservists "twice the citizen" for their willingness to sacrifice their domestic comforts in time of need.

Whether you intend to enlist or apply for a commission, anything you can do to get leadership experience will serve you well when you get to the Navy. Becoming a supervisor or assistant manager at a fast-food place or retail store will teach you a lot about what it takes to lead other people. So will taking on leadership positions in high school clubs, sports teams and local civic associations. The UCMJ invests military leaders with enormous responsibility for the people they lead. Captains of ships have been relieved of duty resulting from mistakes made by petty officers they hardly knew. In the military world, leaders are ultimately responsible for everything everyone under their command does.   The military is one of the few employers that requires its employees to maintain a minimum level of fitness at all times. In fact, the leading cause of involuntary separation from the Navy (firing) is the inability to maintain the proper height-weight standard. If you are fit you will look better in your uniform and feel better among your peers. You will also be ready for those once-in-a-while field exercises that will push your body to the limits.

GETTING STARTED

IF YOU'RE READY TO JOIN THE NAVY THERE IS REALLY ONLY one step to take: Talk to a recruiter. Requirements and opportunities change every day. Websites cannot keep up with new information. The only way to get the answers you need is to go directly to the source.

A recruiter will be able to get you started on the right track. First of all, a recruiter will ask you about your background, about what you have done and what you think you'd like to do for the Navy. A good recruiter will give you all the details you could possibly want. If you decide to move forward you will have to take a test, called the Armed Services Vocational Aptitude Battery, or ASVAB. Your score will determine which ratings you are eligible for. If one appeals to you, you can apply for a billet. If you are accepted you will be slotted into a boot camp division and given a date to report for a physical. It's that simple.

Applying for a commission straight from civilian life is more complicated. You will also need to talk to a recruiter, but in addition, you'll need to complete a stack of paperwork about a mile high, write an essay, and wait for a verdict from a selection board. If you are selected you will be given a date to report for a physical and another date to report to Officer Candidate School.

Keep in mind that simply talking to a recruiter does not incur any obligation on your part. Neither does taking the ASVAB. You don't have to sign anything until you have negotiated the contract you desire. Your contract should spell out all of your initial accession requirements: boot camp, Officer Candidate School, "A" School, "C" School, officer training, etc. If your rating or designator comes with a signing bonus you should get that down on paper too. Once you have the contract you want, sign on the dotted line and get on with it. The rest of your life is calling.

ASSOCIATIONS, PERIODICALS, WEBSITES

" ☐Africa Partnership Station

" www.c6f.navy.mil

" ☐African Center for Strategic Studies

" www.africacenter.org

" ☐American Legion

" www.legion.org

" ☐Canadian Navy

" www.navy.forces.gc.ca

" China Defense Today

" www.sinodefence.com

" ☐Combined Joint Task Force Horn of Africa

" www.hoa.africom.mil

" ☐Defense Link

" www.defenselink.mil

" ☐Defense News

" www.defensenews.com

" ☐French Foreign Legion

" www.legion-recrute.com/en/

" ☐French Navy

" www.defense.gouv.fr/marine_uk

" ☐Global Security

" www.globalsecurity.org

- ☐ Haze Gray and Underway
- www.hazegray.org
- ☐ Indian Navy
- www.indiannavy.nic.in
- ☐ International Journal of Naval History
- www.ijnhonline.org
- ☐ Jane's
- www.janes.com
- ☐ Japanese Maritime Self-Defense Force
- www.mod.go.jp/msdf/formal/english/index.html
- ☐ Marshall Center for European Center for Security Studies
- www.marshallcenter.org
- ☐ Military.com
- www.military.com
- ☐ Naval History and Heritage Command
- www.history.navy.mil
- ☐ Naval Open Source Intelligence
- www.nosi.org
- ☐ Naval Postgraduate School
- www.nps.edu
- ☐ Naval Review
- www.naval-review.org
- ☐ Naval Technology
- www.naval-technology.com

- ☐ Naval War College Review
- www.nwc.navy.mil/press/review/overview.aspx
- ☐ NavSource
- www.navsource.org
- ☐ Navy League of the United States
- www.navyleague.org
- ☐ Navy Times
- www.navytimes.com
- ☐ North Atlantic Treaty Organization
- www.nato.int
- ☐ Royal Australian Navy
- www.navy.gov.au
- ☐ Royal Navy
- www.royalnavy.mod.uk
- ☐ Royal New Zealand Navy
- www.navy.mil.nz
- ☐ Russian Navy (unofficial)
- www.rusnavy.com
- ☐ Sea Waves Magazine
- www.seawaves.com
- ☐ Surface Warfare Magazine
- http://surfwarmag.ahf.nmci.navy.mil
- ☐ Today's Military
- www.todaysmilitary.com
- ☐ Undersea Warfare Magazine

"   www.navy.mil/navydata/cno/n87/mag.html

"   ☐United Nations

"   www.un.org

"   ☐United Nations Security Council

"   www.un.org/docs/sc

"   ☐United States Air Force

"   www.af.mil

"   ☐United States Army

"   www.army.mil

"   ☐United States Coast Guard

"   www.uscg.mil

"   ☐United States Department of Defense

"   www.defenselink.mil

"   ☐United States Department of Veterans Affairs

"   www.va.gov

"   ☐United States Institute of Peace

"   www.usip.org

"   ☐United States Marine Corps

"   www.marines.mil

"   ☐United States Naval Institute

"   www.usni.org

"   ☐United States Navy

"   www.navy.mil

"   ☐Veterans of Foreign Wars

"   www.vfw.org

"   ☐Warship International

"   www.warship.org

"   ☐Warships

"   www.warshipsifr.com

www.ingramcontent.com/pod-product-compliance
Lightning Source LLC
Chambersburg PA
CBHW071020180526
45168CB00003B/1501